WHERE'S THE NARWHAL?

Peter Pauper Press, Inc.
WHITE PLAINS, NEW YORK

First published in the United States by Peter Pauper Press, Inc.
Originally published in the United Kingdom as
Where's the Narwhal? by Orchard Books.

Published by Peter Pauper Press, Inc.
202 Mamaroneck Avenue
White Plains, New York 10601 USA

Library of Congress Control Number: 2020931246

ISBN 978-1-4413-3507-4

Manufactured for Peter Pauper Press, Inc.

Printed in China

7 6 5 4 3 2 1

Visit us at www.peterpauper.com

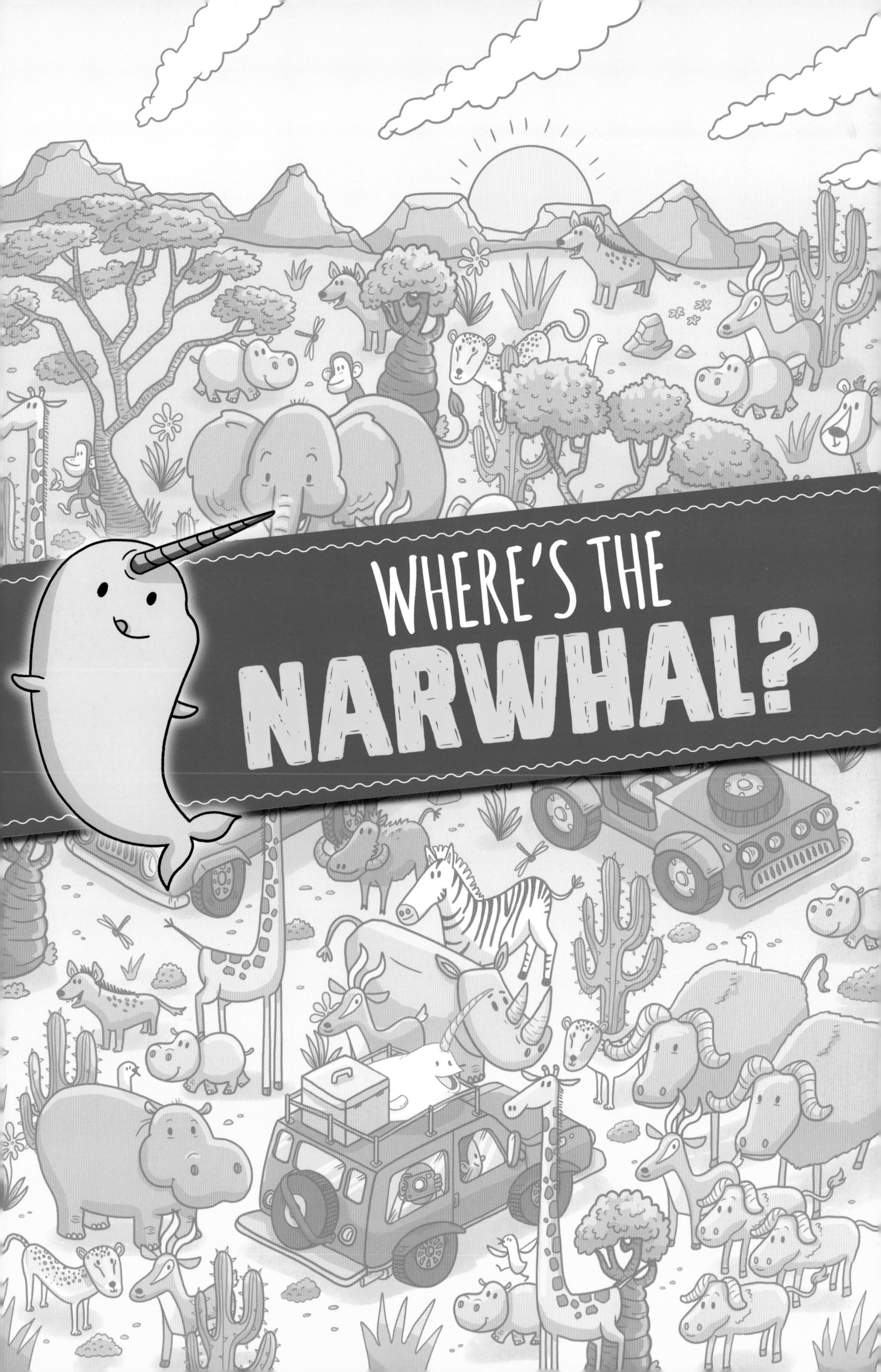
WHERE'S THE
NARWHAL?

MEET THE UNICORNS OF THE SEA!

Narwhals are a type of whale with a special tusk. The tusk is actually a big tooth and can grow up to ten feet long! Narwhals usually live in the Arctic, feeding on fish, squid, and shrimp, but the special narwhals in this book are going on an adventure!

Can you spot the family as they travel around the world?

The answers are at the back of the book, along with some extra things to look out for!

NIALL

Niall looks out for his two younger siblings. He's a natural worrier so the thought of traveling the world is a little scary!

NICOLA

Nicola loves having fun and playing pranks on her family. She is always making mischief wherever she goes!

NOAH

Noah is the baby of the narwhal family and the most excited about their new adventure! He loves exploring his Arctic home but can't wait to travel the world.

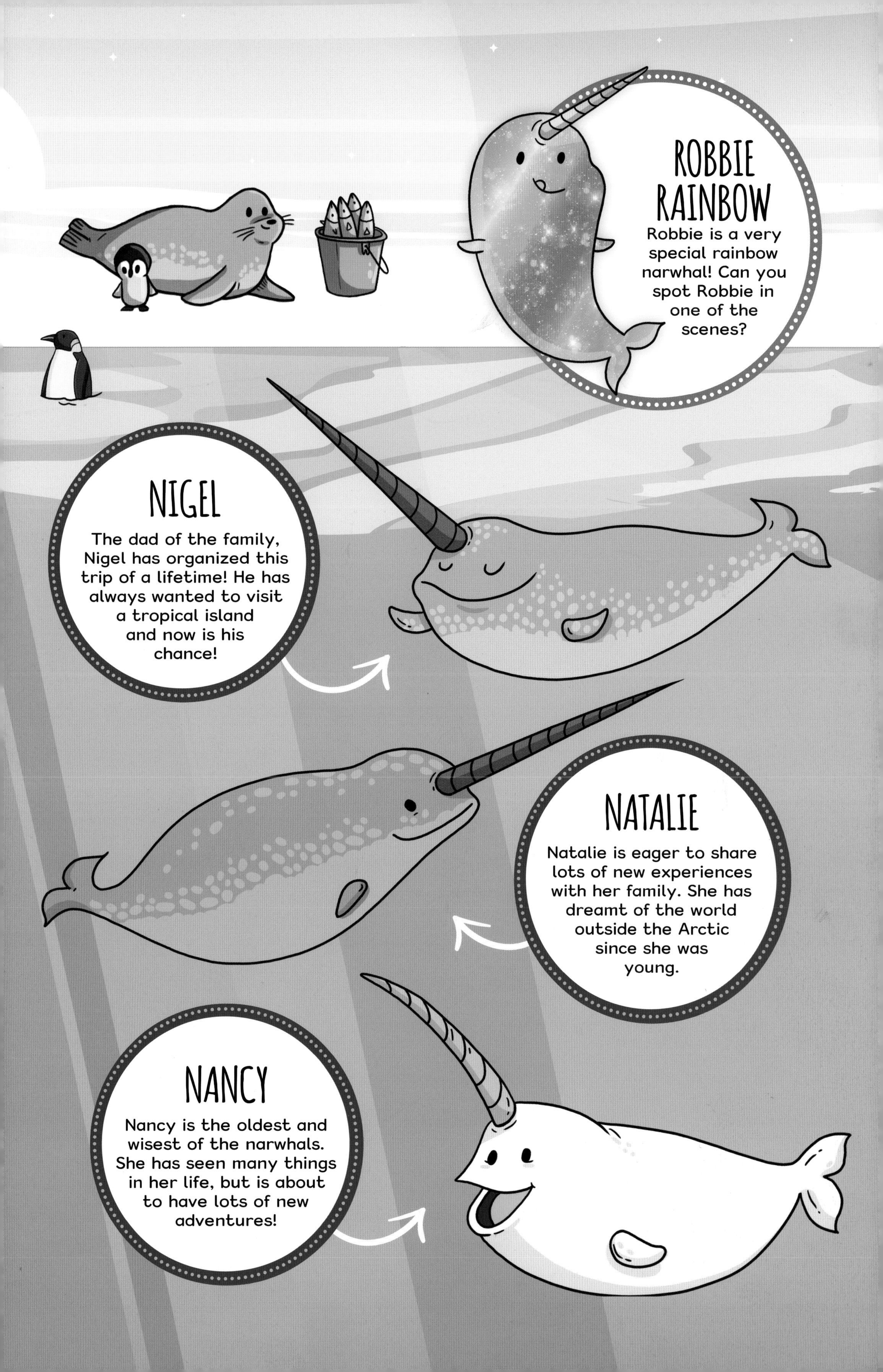
ROBBIE RAINBOW
Robbie is a very special rainbow narwhal! Can you spot Robbie in one of the scenes?
NIGEL
The dad of the family, Nigel has organized this trip of a lifetime! He has always wanted to visit a tropical island and now is his chance!
NATALIE
Natalie is eager to share lots of new experiences with her family. She has dreamt of the world outside the Arctic since she was young.
NANCY
Nancy is the oldest and wisest of the narwhals. She has seen many things in her life, but is about to have lots of new adventures!

SEASIDE
First stop is the beach! The narwhals can't wait to relax in the sun. It's very different than the Arctic!

PARTY TIME
The narwhals have traveled back to the cold, where animals from the Arctic and Antarctic have all gathered for a special party!

PENGUIN PARADE
The narwhals have gotten caught up in a penguin parade!

Odd one out!
Can you
spot the penguin
that looks different
from the rest?

PIER
Down at the town pier, the narwhals are looking forward to a day out. Fresh fish, anyone?

CITY CENTER
The city center is very busy, but the narwhals want to do some shopping. There aren't many shops in the Arctic!

UNDER THE SEA
The narwhals have made friends with lots of orca, beluga, and blue whales.

Odd one out!
Can you spot the whale that looks different from the rest?

MUSEUM
The museum is a great place to learn about the world. The narwhals are particularly excited to see the dinosaur skeletons!

WATER PARK
After all their traveling, the narwhals are relaxing at the water park. They love the slides!

UNICORN GATHERING
The unicorns of the sea have met lots of magical unicorns!

Odd one out!
Can you spot the unicorn that looks different from the rest?

WHITEWATER RAFTING
Time for some thrills! The narwhals are ready for the rapids and whitewater rafting.

THEME PARK
Niall is a bit scared of the rides, but Nicola can't wait to go on the big roller coaster!

CREATURE CRUSH
The narwhals have gotten lost among the giant squid and walruses!

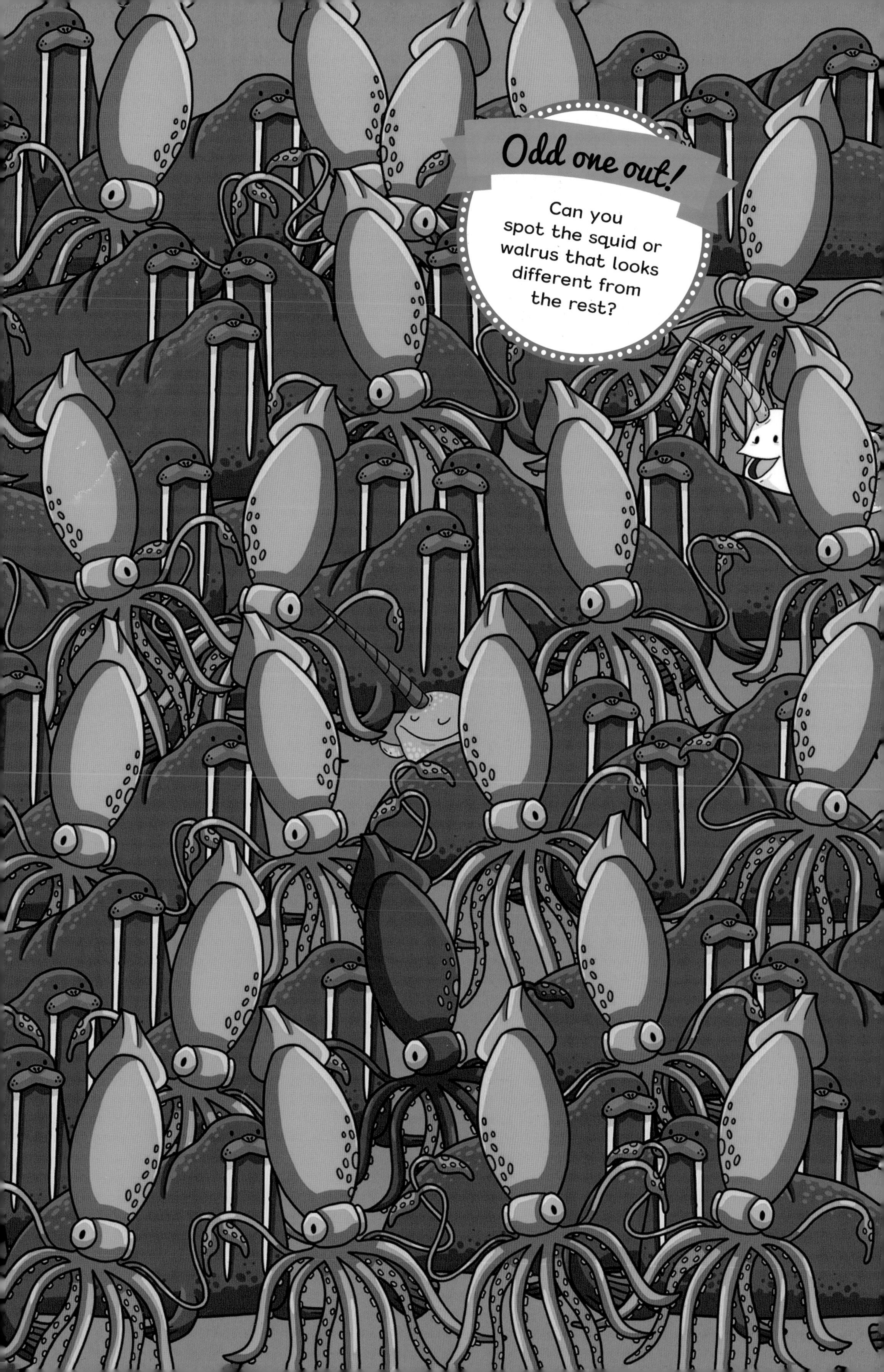

Odd one out!
Can you spot the squid or walrus that looks different from the rest?

SAFARI
The family has traveled all the way to Africa to go on safari. There are so many new animals to see!

AIRPORT
It's time to go home, so the narwhals have come to the airport for their flight. They can't wait to go on their next adventure!

Answers

Now try and find these extra items in every scene!

SEASIDE

- A boy buried in sand ☐
- A dog digging a hole ☐
- A boy eating a hot dog ☐
- A woman with a heart on her hat ☐
- A boy reading a map ☐
- A rhino inflatable ☐
- A woman knitting ☐
- Two children playing paddleball ☐
- A man with a green mohawk ☐
- A man holding two ice creams ☐

PARTY TIME

- An Arctic fox with floaties ☐
- A penguin on a polar bear ☐
- A snowman with a top hat ☐
- A bucket filled with fish ☐
- A penguin jumping off a diving board ☐
- An Arctic hare in an inflatable ring ☐
- A man wearing a yellow jacket ☐
- A polar bear eating a fish ☐
- An Arctic fox fishing ☐
- An Arctic fox carrying a bone ☐

PENGUIN PARADE

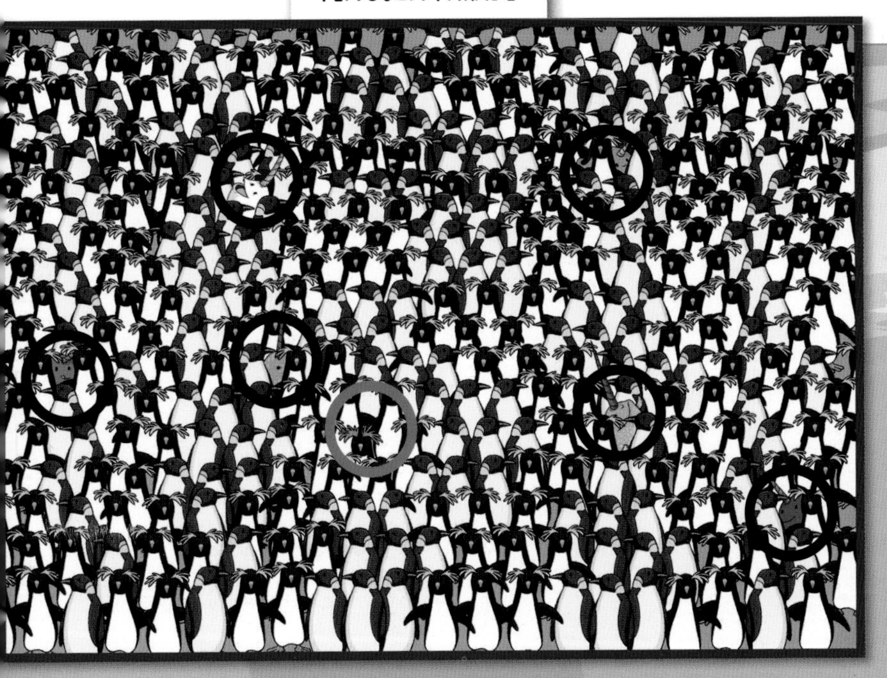

PIER

- Two boys using binoculars ☐
- A seagull eating a french fry ☐
- A baby ☐
- A boy eating cotton candy ☐
- A man in a pirate hat ☐
- A girl with a fish-shaped balloon ☐
- A man reading a newspaper ☐
- A woman eating fish and fries ☐
- A man with a mustache, wearing a tie ☐
- A seagull in a pirate hat ☐

CITY CENTER

- A teddy bear ☐
- A man with a curled mustache ☐
- A man reading a newspaper ☐
- A woman wearing a pink hat ☐
- A man riding a high wheel bicycle ☐
- A dog looking out of a window ☐
- A boy on a scooter ☐
- A tabby cat ☐
- A man wearing a Hawaiian shirt ☐
- A woman looking in a mirror ☐

UNDER THE SEA

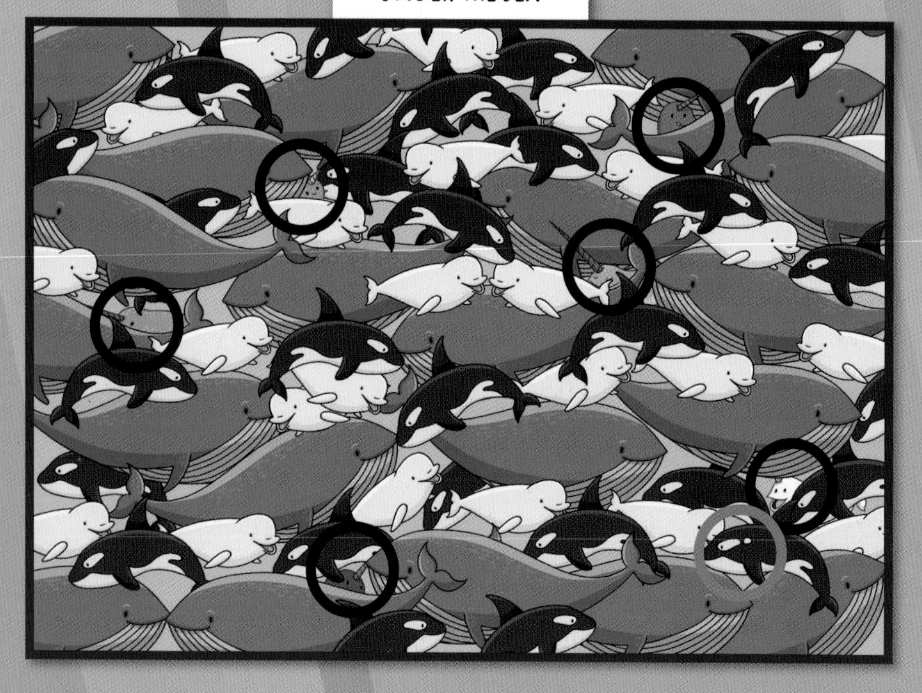

MUSEUM

- A girl wearing orange headphones ☐
- A boy with a spider on his T-shirt ☐
- A girl with a heart on her T-shirt ☐
- A boy with an eye patch ☐
- A man with stripy pants ☐
- A baby playing with blocks ☐
- A boy with a dinosaur on his T-shirt ☐
- A picture of an elephant ☐
- A boy wearing a bow tie ☐
- A man wearing sunglasses ☐

WATER PARK

- A slice of watermelon ☐
- A man taking a shower ☐
- A boy running by the pool ☐
- An alligator inflatable ☐
- Someone wearing flippers ☐
- An orange ball ☐
- A black and white striped towel ☐
- A lifeguard ☐
- Someone climbing out of the pool ☐
- Four starfish inflatables ☐

UNICORN GATHERING

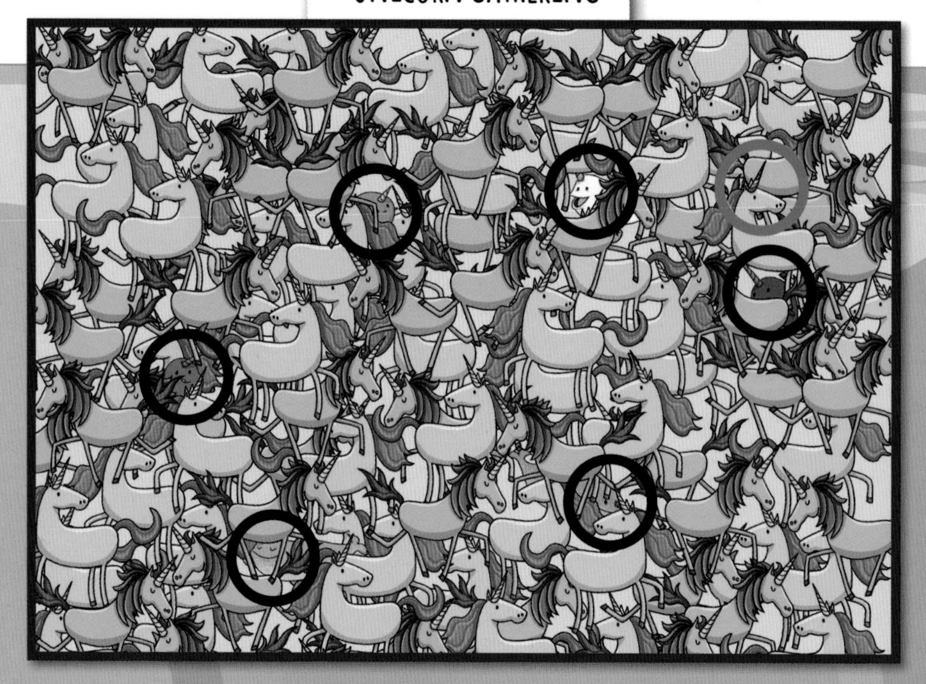

WHITEWATER RAFTING

- A man eating a sandwich ☐
- Two people in the water ☐
- Two pink kayaks ☐
- A man with walking sticks ☐
- A man with a long white beard ☐
- Someone wearing green sandals ☐
- A dog next to a ball ☐
- A man with a backwards cap ☐
- A woman holding a pink cup ☐
- A man with binoculars ☐

THEME PARK

- Someone scared of the teacups ☐
- A clown ☐
- A man on a ride wearing a hat ☐
- A lion balloon ☐
- A girl dressed as a superhero ☐
- A man eating a hot dog ☐
- A Frankenstein's monster ☐
- A girl reading by the ferris wheel ☐
- An octopus ☐
- A mummy ☐

CREATURE CRUSH

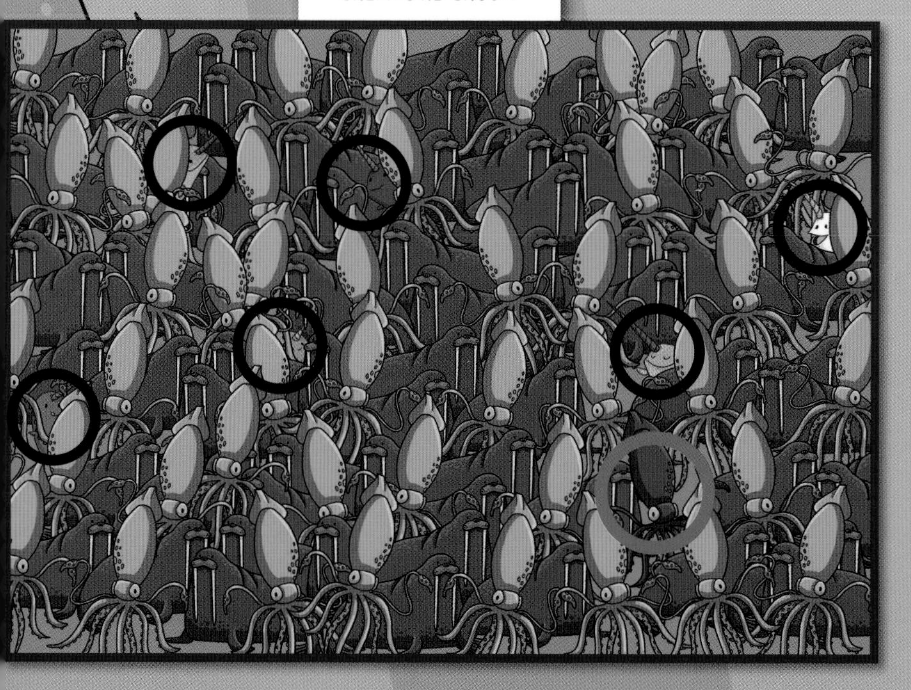

SAFARI

- A lion sitting on a rock ☐
- Two men taking pictures ☐
- A hungry hippo ☐
- A monkey holding a banana ☐
- Two sad frogs ☐
- A pair of dice ☐
- Spotlights ☐
- A boy enjoying the safari ☐
- A monkey climbing a tree ☐
- A lion taking a bath ☐

AIRPORT

- A man pointing at the escalator ☐
- Find six baby carriages ☐
- A palm tree ☐
- A boy with lightning on his T-shirt ☐
- A man wearing a straw hat ☐
- A muffin ☐
- A boy with a green mohawk ☐
- Two babies ☐
- A woman with blue hair ☐
- A boy crying ☐